DeltaScience ContentReaders™

Changes in Ecosystems

Contents

Preview the Book 2
How Do Ecosystems Work? 3
   About Ecosystems 4
   Needs of Living Things 6

Cause and Effect 8
Why Do Ecosystems Change? 9
   Natural Events Cause Changes 10
   Living Things Cause Changes 11

Main Idea and Details 16
What Happens When Ecosystems Change? .. 17
   Succession 18
   Survival and Extinction 19

Compare and Contrast 20
How Can People Help Protect Ecosystems? .. 21
   Restoring and Protecting Ecosystems 22

Glossary 24

Build Reading Skills
Preview the Book

You read nonfiction books like this one to learn about new ideas. Be sure to look through, or *preview*, the book before you start to read.

First, look at the title, front cover, and table of contents. What do you guess you will read about? Think about what you already know about ecosystems.

Next, look through the book page by page. Read the headings and the words in bold type. Look at the pictures and captions. Notice that each new part of the book starts with a big photograph. What other special features do you find in the book?

Headings, captions, and other features of nonfiction books are like road signs. They can help you find your way through new information. Now you are ready to read!

How Do Ecosystems Work?

MAKE A CONNECTION
Pelicans, elephants, and grasses are living things in this African grassland ecosystem. Soil, water, and air are nonliving things in the ecosystem. How do you think these living and nonliving things affect one another?

FIND OUT ABOUT
- the parts of an ecosystem
- the needs of living things

VOCABULARY
habitat, p. 4	niche, p. 5
ecosystem, p. 4	population, p. 5
organism, p. 4	community, p. 5
climate, p. 4	environment, p. 6
species, p. 5	competition, p. 7

About Ecosystems

A desert coyote has places where it sleeps, hunts, and finds water. These places are all part of its home, or **habitat**. Many animals and plants live in a desert. A desert is an example of an ecosystem. Earth has many different kinds of ecosystems.

An **ecosystem** includes all the living and nonliving things in one place. Plants and animals are living things. Living things are sometimes called **organisms**. Rocks, soil, sunlight, and water are nonliving things. Climate is another nonliving part of an ecosystem. **Climate** is the average weather in an area over many years.

An ecosystem also includes the interactions among its living and nonliving things. When things interact, they affect one another. For example, the kind of soil in a desert affects the plants that grow there.

▲ A burrowing owl's habitat includes its hole in the ground, or burrow.

▲ Plants, animals, soil, and a dry climate are all part of this desert ecosystem. So are their interactions.

▲ A population of coyotes may be part of a desert community.

Scientists use certain words when they talk about the living things in an ecosystem. Living things of the same kind belong to the same **species**. For example, burrowing owls and great horned owls are two different species.

Many different species can share the same habitat. But each organism has its own role, or **niche**. For example, an antelope jackrabbit's niche might include feeding on grasses and becoming food for a coyote.

All the members of one species living in one place make up a **population**. One part of a desert might have a population of five coyotes, for example.

Together, all the populations in one place make up a **community**. A desert community can include populations of many different species of animals and plants. All the different populations in a community interact.

✓ An ecosystem includes all the living and the nonliving things in one place. What else does an ecosystem include?

Needs of Living Things

All organisms need certain things to stay alive. Animals need air, water, food, shelter, and space to live. Plants need air, water, light, and space to live. Plants also need nutrients such as the minerals in soil.

Resources are things that an organism can use to meet its needs. Resources come from an organism's environment. An **environment** is all the physical things and conditions around an organism. An elk may eat grasses and drink water from a river in its environment. These things are resources for the elk. They help it meet its needs.

Each environment has only so many resources. Some organisms are better than others at getting or using those resources. These organisms are the ones that stay alive.

This elk is meeting its need for food by eating the grass in its environment. Grass is a resource for the elk.

◀ Grizzly bears sometimes compete for fish to eat.

A living thing is affected by the
- kinds of other living things in its environment
- number of other living things in its environment

Many different organisms often depend on the same resources. This can cause a kind of struggle. The struggle for resources is called **competition**. Competition can happen among individuals of the same species. Competition also can happen among different species.

A healthy ecosystem is in balance. A small change in one part can upset that balance. The whole ecosystem may be affected. For example, grizzly bears may eat fish from a river. But suppose there are fewer fish in the river. There might not be enough food for all the bears. Some bears might have to find other things to eat. Then they might be competing for food with different animals. This could affect whether those other animals can find enough to eat.

 How can competition for resources affect a living thing?

REFLECT ON READING
You previewed pictures, captions, and other book features before reading. Which features helped you the most as you read about ecosystems? How did they help?

APPLY SCIENCE CONCEPTS
Think about a potted plant you have seen growing indoors. What needs does that plant have? How are those needs met? Talk about your ideas with the class.

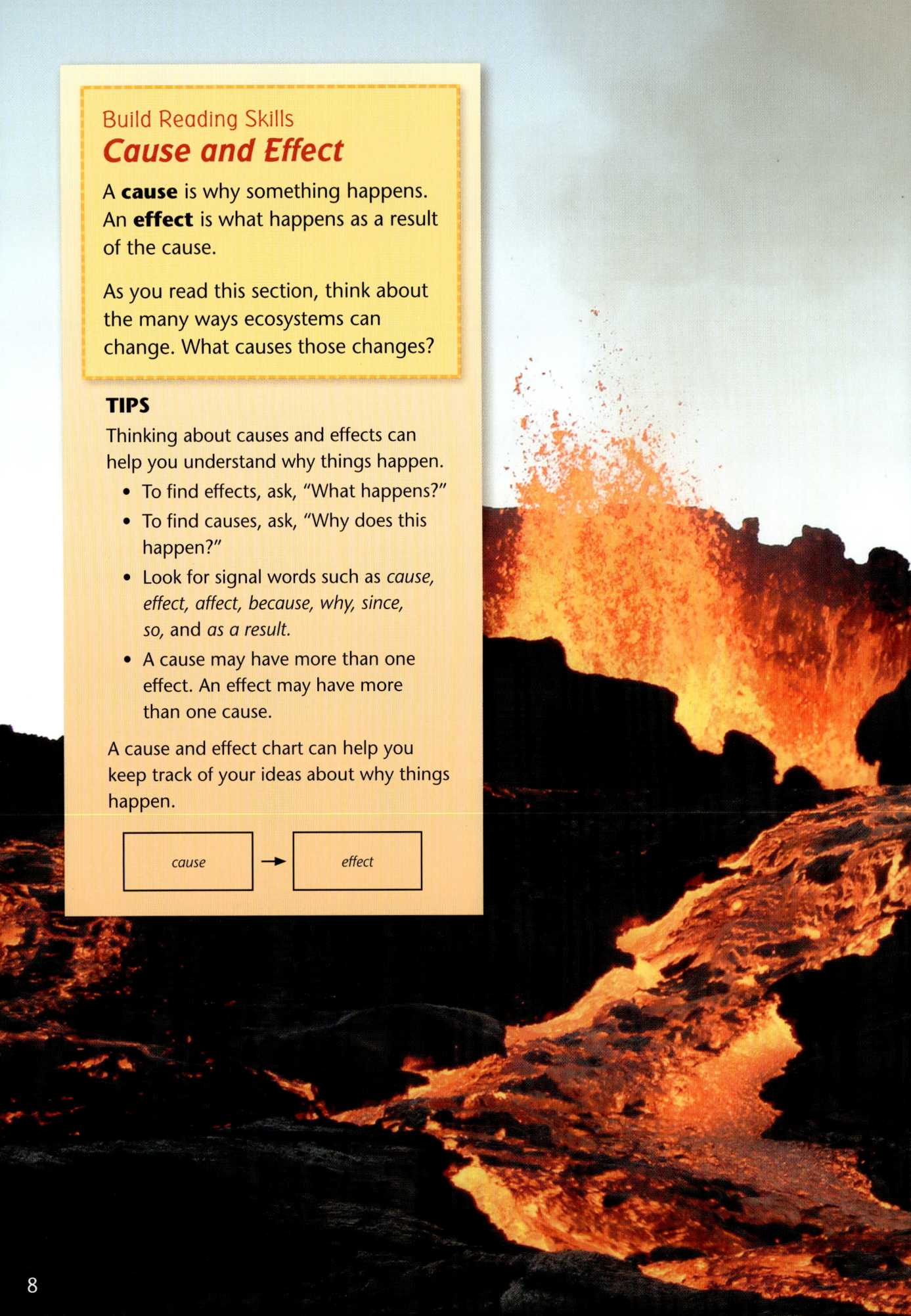

Build Reading Skills
Cause and Effect

A **cause** is why something happens. An **effect** is what happens as a result of the cause.

As you read this section, think about the many ways ecosystems can change. What causes those changes?

TIPS

Thinking about causes and effects can help you understand why things happen.

- To find effects, ask, "What happens?"
- To find causes, ask, "Why does this happen?"
- Look for signal words such as *cause, effect, affect, because, why, since, so,* and *as a result.*
- A cause may have more than one effect. An effect may have more than one cause.

A cause and effect chart can help you keep track of your ideas about why things happen.

| cause | → | effect |

Why Do Ecosystems Change?

MAKE A CONNECTION
Some places have very active volcanoes. What changes might happen in an ecosystem when a volcano erupts?

FIND OUT ABOUT
- ways that events in nature change ecosystems
- ways that living things change ecosystems

VOCABULARY
pollution, p. 14
acid rain, p. 15

Natural Events Cause Changes

Natural events can affect ecosystems. Storms can cause floods, and winds can blow down trees. An ecosystem may go through a long period of dry weather called a drought. Some of the plants that grow there may die because they do not have enough water. Then the animals that eat those plants would compete for any plants that are left. Earthquakes, volcanoes, and forest fires are other natural events that can change ecosystems.

Changes in ecosystems can be harmful to some living things. But they can be helpful to others. Suppose lightning starts a forest fire that burns the trees in an area. Tall trees no longer block the sunlight. So new plants that sprout can get enough sunlight to grow. The ash from the fire adds nutrients to the soil. This also helps the new plants.

 Name four natural events that can change ecosystems.

A forest fire can change the soil and the kinds of living things in an ecosystem. ▼

▲ Beavers change ecosystems by building dams across streams.

Living Things Cause Changes

All living things cause changes in their environments. These changes can affect whole ecosystems. Suppose beavers build a dam across a stream. The dam blocks the flow of water. Then a pond forms. The pond is a safe place to live for beavers and other wetland organisms. But what about the other animals and plants that used to live there? Their woodland environment is now flooded. The new wetland environment might not meet their needs.

Plants also can cause changes that affect ecosystems. Think of a field that was once farmland. It is a sunny environment. Plants such as grasses grow well there. Nutrients are added to the soil when older grass plants die and decay. Then pine trees might grow in the soil. Over time, the trees get bigger. More trees grow. The sunny field ecosystem changes to a shady forest ecosystem.

Disease in plants or animals also can cause changes in ecosystems.

Humans make changes in many ecosystems. For example, people sometimes bring a plant or an animal to a place where it does not usually live. These plants or animals are called *introduced species*. Introduced species can harm other organisms. Introduced species also can compete with other organisms for resources. The new habitat may not have an animal that eats the introduced species. If so, the introduced species can multiply quickly.

A European insect called the gypsy moth was brought into the United States in about 1869. It was thought that the insect could be used for making silk. At first the moths were kept in one place. But some moths got away. Now gypsy moths have spread across the northeastern United States and beyond. Gypsy moth caterpillars have harmed many trees by eating their leaves.

The gypsy moth is an introduced species in the United States. Gypsy moth caterpillars have harmed millions of acres of trees.

A busy road cuts through this tortoise's habitat. ▶

People also change ecosystems by using or developing the land. We often clear land for new buildings. We also build highways, bridges, and dams. Raising animals, growing crops, and digging mines change the land, too.

Land development can make habitats smaller. Plants and animals in developed areas may have less space, food, and water. They may no longer find what they need to stay alive. As a result, these plants and animals may move to other areas. This may cause more competition in those places.

People also use land for getting rid of trash. More than half our solid waste is put in landfills. Landfills take up space that was once a home for plants and animals. People must design landfills carefully. Otherwise, wastes can leak and harm nearby habitats.

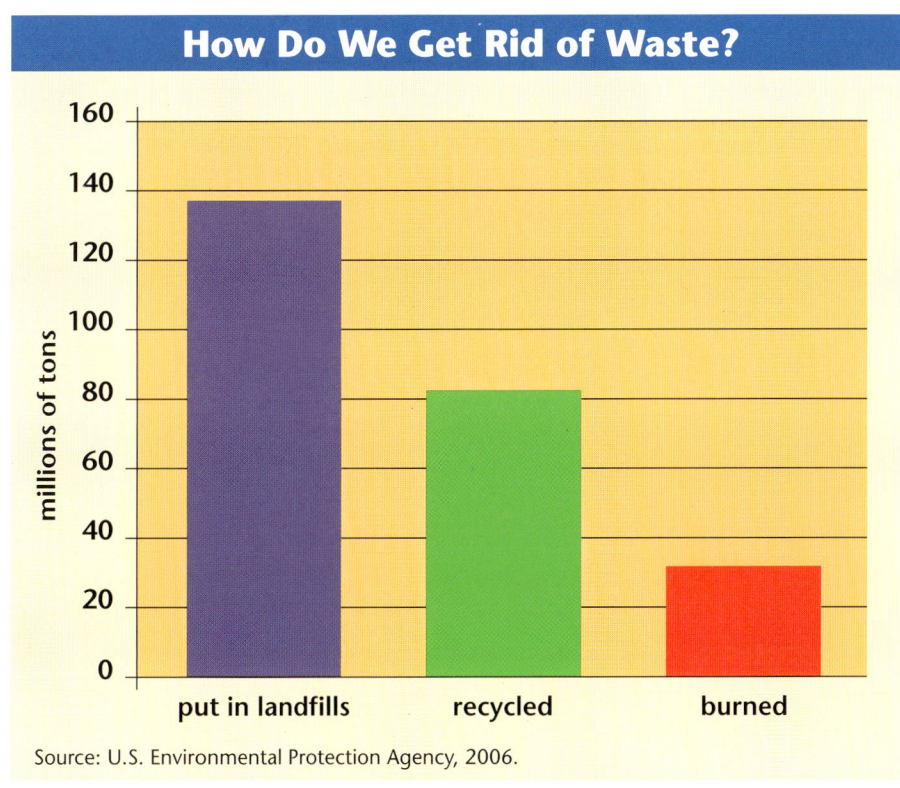

◀ People in the United States got rid of about 251 million tons of solid waste in 2006. This graph shows how much of that waste was put in landfills, recycled, and burned.

Source: U.S. Environmental Protection Agency, 2006.

People can damage ecosystems by putting harmful substances into the soil, water, or air. This is called **pollution**. Pollution can make food and water unsafe. It can kill plants and animals. For example, sometimes oil carried by a ship spills into the ocean. The oil pollutes the water and harms wildlife.

Another kind of pollution can happen when people use harmful chemicals. For example, some chemicals are used to kill weeds or harmful insects. But these chemicals also can harm helpful organisms like birds, worms, and bacteria.

People sometimes add fertilizers to soil to help plants grow. But when it rains, fertilizers can wash into ponds and lakes. The fertilizers help tiny water organisms called algae grow quickly. When the algae die, bacteria break them down. These bacteria use up most of the oxygen in the water. As a result, fish die because they cannot get the oxygen they need.

▲ Birds and other wildlife can be harmed or killed by pollution from an oil spill.

▲ Algae can grow too quickly when fertilizers wash into ponds.

▲ Some power plants and factories burn fossil fuels such as coal and oil. Burning these fuels can give off waste gases that can pollute the air.

Other kinds of pollution are a result of burning fossil fuels such as coal and oil. Many power plants burn coal to make electricity. Cars burn fuels made from oil. But burning fossil fuels can give off gases that can pollute the air.

Burning fossil fuels also can cause a kind of pollution called **acid rain**. The burning fuels give off substances that mix with water vapor in the air. This forms acid rain. Acid rain can cause changes in lakes and streams. These changes can harm plants and animals there. Acid rain also can damage buildings and statues.

 How do humans change ecosystems? Tell about four ways.

REFLECT ON READING
Make a cause and effect chart like the one on page 8. Write "ecosystems change" in the effect box. Think of some possible causes of changes in ecosystems. Add them to the chart.

APPLY SCIENCE CONCEPTS
Think about the ecosystem you live in. How have people changed it? Write about the changes in your science notebook. How do you think the changes have affected other living things?

Build Reading Skills
Main Idea and Details

The **main idea** of a paragraph or part of a book is the most important point. **Details** give more information about the main idea.

As you read this section, look for the main idea about adaptations.

TIPS

The topic sentence tells the main idea of a paragraph. It is often the first sentence in the paragraph. To find the main idea, ask, "What is this paragraph mostly about?"

Details may answer Who, What, When, Where, Why, and How questions about the main idea. Details can be
- examples
- descriptions
- reasons
- other facts

A concept web can help you keep track of the main idea and details.

What Happens When Ecosystems Change?

MAKE A CONNECTION
Wild bison used to live in many parts of North America. How has the bison's ecosystem changed?

FIND OUT ABOUT
- how the populations in an ecosystem can change over time
- ways living things stay alive in a changing ecosystem
- what can happen to living things when an ecosystem no longer meets their needs

VOCABULARY
succession, p. 18
adaptation, p. 19
extinct, p. 19
endangered, p. 19
fossil, p. 19

Succession

Ecosystems change all the time. You have learned that when an ecosystem changes, some species may no longer get what they need. As a result, those species move away or die. But the changed ecosystem may have just what some other species need. Those species move in. A gradual change in the kinds of organisms living in an ecosystem is called **succession**.

Remember how a field that was once farmland can change into a forest. That is an example of succession. The grasses in the field are gradually replaced by shrubs and trees. The kinds of animals living there also change.

Succession also can happen after a sudden event greatly changes an ecosystem. Mount St. Helens is a volcano in the state of Washington. Its eruption in 1980 killed thousands of plants and animals. But within a year, seeds blew into the area and started to grow. In some places, shrubs had stayed alive. Some animals also lived through the eruption. Other animals moved in as plants started to spread.

 What is succession?

Wildflowers were some of the first plants to grow after Mount St. Helens erupted. Scientists are still studying the area around the volcano to learn more about succession. ▶

18

▲ A polar bear has thick fur. This adaptation helps the bear live in a very cold environment.

▲ The dodo is an extinct bird species.

Survival and Extinction

Sometimes, special body parts or behaviors help organisms stay alive when their environments change. The organisms pass these helpful features on to their young. As a result, over many generations, the whole species changes, or adapts. Features that help organisms survive are called **adaptations**. A polar bear's thick fur helps it survive in a very cold environment. Some fish swim in large groups. This adaptation makes it harder for bigger fish to catch and eat one of them.

Sometimes a species cannot adapt to changes in its environment. As a result, that species may die out. A species is **extinct** when all its members are dead. A species is **endangered** when it is close to becoming extinct. We can learn about extinct species from fossils. **Fossils** are the preserved remains of once-living things.

 Give an example of an adaptation. Tell how it helps an organism survive.

REFLECT ON READING
Make a concept web like the one on page 16. Use the web to keep track of ideas about adaptations. Put the main idea in the middle. Then add details, such as examples.

APPLY SCIENCE CONCEPTS
Find out about an animal that is endangered. Use books, magazines, or the Internet. Talk with a partner about what people can do to save the animal.

Build Reading Skills
Compare and Contrast

When we **compare** two things, we tell how they are alike. When we **contrast** two things, we tell how they are different.

As you read this section, think about how the three Rs, reduce, reuse, and recycle, are alike and different.

TIPS

Compare and contrast by following these steps.

- Choose two related things.
- To compare them, ask, "How are they alike?"
- To contrast them, ask, "How are they different?"

A Venn diagram can help you compare and contrast.

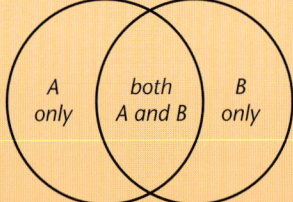

How Can People Help Protect Ecosystems?

MAKE A CONNECTION
Some people reuse bags when they go to the grocery store. They don't use new paper or plastic bags each time. How do you think this is helpful?

FIND OUT ABOUT
- ways people are helping to return damaged ecosystems to their natural state
- ways people are protecting ecosystems

VOCABULARY
conservation, p. 22

Restoring and Protecting Ecosystems

People cause many changes to ecosystems. Some of these changes can be helpful. Others can be harmful for us and for other living things. But people are also helping to return, or restore, ecosystems to their natural state. This is called *reclamation*. Picking up trash on a beach is reclamation. So is planting new trees where a forest has been cut down.

Many countries have laws that protect ecosystems. Some laws help control pollution. Others help protect endangered species.

Conservation is the wise use of natural resources. Conservation can help protect ecosystems. People practice conservation by following the three Rs: reduce, reuse, and recycle. These things save resources and reduce pollution.

▲ People can plant trees to help fix, or reclaim, land that has been logged or mined.

▲ Taking a train or bus uses less fuel than taking separate cars. This helps save resources and reduces pollution.

▶ Plastic, glass, paper, and metal can be recycled. Recycling saves resources and space in landfills.

Reduce means to use less of something. Using less electricity conserves the resources used to make it. People can use less electricity by turning off the lights and TV when they are not being used. People can reduce waste by using products that are refillable. People also can reduce their use of harmful chemicals.

Reuse means to use again. People can reuse containers instead of throwing them away. Broken objects can be fixed instead of replaced. We can give outgrown clothes to others.

Recycle means to turn old products into material for new products. For example, the plastic in many bottles can be recycled. It can be made into new bottles, carpeting, or cloth. We also can recycle metal, glass, and paper.

 Tell one way people can help restore a damaged ecosystem. Then tell one way people can protect ecosystems.

REFLECT ON READING
Make a Venn diagram like the one on page 20. Choose two of the three Rs (reduce, reuse, and recycle). Use the diagram to show how they are alike and different.

APPLY SCIENCE CONCEPTS
Think about plastic packaging, toys, and other items in your home. What are some ways your family can use less plastic? List your ideas in your science notebook.

23

Glossary

acid rain (AS-id RAYN) a kind of rain that forms when chemicals from air pollution mix with water vapor in the air **(15)**

adaptation (ad-ap-TAY-shuhn) a body part or behavior that helps an organism survive in its environment **(19)**

climate (KLYE-mit) the average weather in an area over many years **(4)**

community (kuh-MYOO-nuh-tee) all the different populations of organisms living together in one place **(5)**

competition (kom-puh-TISH-uhn) the struggle among living things that share the same resources such as sources of food, water, or shelter **(7)**

conservation (kon-sur-VAY-shuhn) the wise use of natural resources **(22)**

ecosystem (EE-koh-sis-tuhm) all the living and nonliving things in one place and all their interactions; examples are grassland, desert, rain forest, and freshwater ecosystems **(4)**

endangered (en-DAYN-jurd) when a species is close to becoming extinct **(19)**

environment (en-VYE-ruhn-muhnt) all the physical things and conditions, such as soil, air, climate, plants, and animals, that surround a living thing **(6)**

extinct (ik-STINKT) when a species has died out and there are no longer any of its kind alive on Earth **(19)**

fossil (FOS-uhl) the preserved remains or traces of a once-living thing **(19)**

habitat (HAB-i-tat) the place in nature that is home to a living thing **(4)**

niche (NICH) the role of a living thing in its habitat; includes where an organism lives, what it eats or takes in from its surroundings, and how it affects the other living and nonliving things in its habitat **(5)**

organism (OR-guh-niz-uhm) a living thing such as a plant or an animal **(4)**

pollution (puh-LOO-shuhn) the act of putting harmful materials into the air, soil, or water **(14)**

population (pop-yuh-LAY-shuhn) all the members of one species living in a place **(5)**

species (SPEE-sheez) a group made up of all the living things of the same kind; burrowing owls and great horned owls are two different species of birds **(5)**

succession (suhk-SESH-uhn) a gradual change in the kinds of organisms living in an ecosystem **(18)**